Ham Radio Activities

The Complete Amateur Radio Contesting Manual

Tips & Techniques for Competing & Winning in a Ham Radio Contest

Introduction

I want to thank you and congratulate you for downloading the book, "*Ham Radio Activities: The Complete Amateur Radio Contesting Manual - Tips & Techniques in a Ham Radio Contest*".

This book is the ultimate amateur radio-contesting manual.

"Ham" that is the name given to those who operate ham radios. Ham radio contest, on the other hand, are a different ball game altogether.

In this guide, we shall be discussing how although the contest aspect of ham radio operation intimidates many, at the heart of the contest is the need for connection and fun.

This guide shall equip you with the requisite knowledge you need to participate in ham radio contests with skill and confidence. After learning the different types of contests, how they work and their rules, the software and hardware to use, you will start building up your QSO count and adding interesting DX contacts to your list of contacts.

Let's begin.

Thanks again for downloading this book. I hope you enjoy it!

© Copyright 2018 by _____ - All rights reserved.

This document is geared towards providing exact and reliable information in regards to the topic and issue covered. The publication is sold with the idea that the publisher is not required to render accounting, officially permitted, or otherwise, qualified services. If advice is necessary, legal or professional, a practiced individual in the profession should be ordered.

- From a Declaration of Principles which was accepted and approved equally by a Committee of the American Bar Association and a Committee of Publishers and Associations.

In no way is it legal to reproduce, duplicate, or transmit any part of this document in either electronic means or in printed format. Recording of this publication is strictly prohibited and any storage of this document is not allowed unless with written permission from the publisher. All rights reserved.

The information provided herein is stated to be truthful and consistent, in that any liability, in terms of inattention or otherwise, by any usage or abuse of any policies, processes, or directions contained within is the solitary and utter responsibility of the recipient reader.

Under no circumstances will any legal responsibility or blame be held against the publisher for any reparation, damages, or monetary loss due to the information herein, either directly or indirectly.

Respective authors own all copyrights not held by the publisher.

The information herein is offered for informational purposes solely, and is universal as so. The presentation of the information is without contract or any type of guarantee assurance.

The trademarks that are used are without any consent, and the publication of the trademark is without permission or backing by the trademark owner. All trademarks and brands within this book are for clarifying purposes only and are the owned by the owners themselves, not affiliated with this document.

Table of Contents

Introduction

Ham Radios Basics: Understanding Amateur Radio

    About Propagation

    About DX-Ing, Contests, and Awards

Contesting 101: A Cultural History of Amateur Radio Contesting

    About Contests

    Brief History of Amateur Radio Contesting

Getting Started: Getting Your Ham Radio License

    About FCC Licensing

    Licensing: About Frequency Privileges

    License Classes and their Various Transmission Privileges

    About The License Examination

    How to Prepare for the License Exam

Amateur Radio Competition Equipment

Making Your First QSO: A Simple, Ham Radio Contest Guide to Making Contacts

The Main Amateur Radio Contests and Their Rules

   ARRL DX Contest

   CQ World Wide DX Contest

   ARRL "Field Day"

Contesting Tips and Techniques for Better Amateur Radio Contesting

Conclusion

To get started, we shall go all the way back as a way to help you understand ham radios and their history.

Ham Radios Basics: Understanding Amateur Radio

Also called amateur radio, Ham radio is the use of select radio frequencies for communication. The communication can be private, emergency, non-commercial, or even for experimentation purposes.

The term "amateur" (as in Amateur Radio) refers to persons interested in radio technology for personal reasons (without any monetary interests) and to differentiate ham radio broadcasting from commercial broadcasting, professional two-way radio (such as those used by taxis, aviation, and marine), and public safety communication such as those used by the police and fire marshals. On the one hand, radio sport (also called contesting), is the competitive aspect of amateur radio where amateur radio operators (hams) compete against each other. An amateur radio operator is a person who uses radio equipment within an amateur radio station that has designated radio frequencies that have been assigned to an amateur radio service to engage other hams (amateur radio operators) in two-way personal communications.

To transmit on a specific radio frequency, amateur radio operators ought to pass an examination administered by a governmental authority, an examination that tests their grasp of applicable regulations, radio theory, electronics, and radio operation. Upon passing the exam, the hams get an amateur radio license.

One component of the license granted after passing the regulatory exam is a call sign that hams use for identifying themselves whenever they are communicating with other hams. According to the International Amateur Radio Union, they are over 2.6 million amateur radio operators spread around the world.

In a ham radio contest, hams, in teams or as individuals, use an amateur radio station (and its assigned frequencies) to contact, and exchange information with as many other amateur radio stations (other hams) within a specified time.

Each ham radio contest has its own rules that define the radio bands, the mode of communication used, as well as the kind of information exchanged during the contact. The contacts contacted throughout the said session make up a score that determines the ranking of each of the participating station.

As we have hinted at many times, hams use specific radio frequencies for communication. Because of the variety of frequencies, hams can use many frequency bands on the radio spectrum. In the U.S., the Federal Communications Commission (FCC) is responsible for allocating frequencies for amateur radio use.

For specificity purposes, hams can operate above the AM broadcast band and gigahertz range with many hams operating in the frequency range of AM radio band (1.6MHz) to slightly above the citizen band of 27MHz. To listen in to an amateur radio communication, non-hams can use radio scanners and receivers.

In the day, the best frequencies for long-distance communications are 15-27MHz. At night, the best frequency ranges for communication are 1.6-15MHz. From a historical perspective, bands within this range are what we call short-wave band (short-wave radio) and as you can guess, they are different from the frequencies used by most modern TV and radio stations.

Short-wave frequencies operate from a line-of-sight. Because of this, and because short waves bounce off the ionosphere from the transmitter to the antenna of the receiver, they have a limitation of 40-50 miles (the higher the frequency, the shorter the wavelength).

Other than using voice, some ham radio operators choose to use Morse code. As you may know, Morse code signals are more reliable and can get through when voice transmissions fail. If you choose to integrate the Morse code element into your ham radio learning experience, you will have to [learn Morse code](#).

As we indicated earlier, Ham radio broadcast are different from normal radio broadcast. One of the key differences is that although ham radios can indeed broadcast in all directions (as a normal radio stations do when a DJ broadcasts a transmission to many tuned in listeners), ham radio communications are often two-way communications with other individual hams or in a roundtable setting of hams in a group.

Because of the short-range nature of the short wave, roundtable ham communications may be in the same town, state, country, continent, or even consist of a mixture of hams drawn from different countries depending on the time of day and the frequency in use.

At predetermined frequencies and times, hams also engage in networks called nets where they exchange third-party information and messages. For instance, when there are emergencies, hams exchange specific information such as health and welfare information.

When it comes to the equipment used by hams, hams use a variety of equipment some of which include radios and antennas with some hams choosing to use [radioteletype (RTTY)](#) that eliminate the use of teletype with computer screens.

In addition to using radios, hams also use telegraph equipment, cameras, laser, computers, and in the case of hams that offer professional services—such as those who offer emergency communication—their own satellites.

To the casual onlooker, the radios used by hams may seem very basic. While they are, they also use very sophisticated technologies (this should not dissuade you from engaging in the ham contests or being a ham enthusiast). In fact, many enthusiasts have varied background with some being experts in the field of technology while others know next to nothing. Think of operating a ham radio in the same way you think of computers: many can use it but very few know about how it works.

Some hams—especially those with a technology background—choose to create their own ham radio stations using equipment they design themselves. However, many hams choose to create ham stations using factory-built equipment that is readily available online and at stores countrywide. "Home brewing" is the name given to the process setting up a ham station and is one of the most enjoyable aspects of the hobby.

About Propagation

Propagation is one of the most important concepts in ham radios. The term refers to the process of the travel of radio waves through the air as they bounce from safe to surface as they make their way to a radio antenna.

Hams are constantly monitoring the atmosphere for atmospheric conditions such as storms, solar flares and other conditions that may affect the propagation or transmission of radio signals from receiver to antenna simply because the weather can affect the effectiveness of transmissions. This makes having the right equipment (especially an antenna because it's responsible for receiving radio signals) a requisite. A radio antenna can be complex or simple; it can be a massive tower or it can be something as simple as a wire attached to the circuitry of a radio.

While voice, Morse, and in the modern age, digital operations—most ham stations in existence today combine computers and radio equipment—and ham stations are the most commonly used types of ham radios and communication, from an operational perspective, most hams start out using VHF FM, battery-operated handheld transceivers that receive on one frequency and transmit on another.

To use these FM repeaters, ham enthusiasts have to join local radio clubs that set up and support them by borrowing antenna space from TV station towers (with permission of course) can re-broadcast received signals further thus extending the range.

For this purpose, the FM repeaters receive a signal at a time and immediately rebroadcast it to another frequency that is more powerful than what is available to the hand-held radio. This has the effect of extending the signal range of the hand-held radio. Many developed countries have thousands of FM repeaters and whenever you travel—or any ham radio enthusiast travels—you can find a repeater to use for communication.

Most of the available repeaters use common transmit and receive frequency pairs informally assigned by groups of hams so that any of the frequencies in use does not encroach on the frequency of another repeater and thus cause unwanted interference.

Other than hand-held transceivers and FM repeaters, ham radio enthusiasts also use amateur radio satellites. In this case, ham radio enthusiasts use a satellite overhead to communicate through their hand-held radios. This is very effective because as you may know, natural weather occurrences such as hurricanes or tornadoes tend to disrupt cellphone and telephone communication. In such instances, ham radios come in handy for emergencies. One British satellite repeater has an uplink (receiver) at 145.975 MHZ and a simultaneous downlink (rebroadcast) at 435.070 MHZ.

"Chewing the rag"

As we discussed earlier, from a contest perspective, the purpose of a ham radio contest is to contact as many people as possible. The communications most common in these conversations has a general reference term, "chewing the rag." When enthusiasts "chew the rag," the contacts they speak to become "ragchews," a term used to refer to a contact or the actual communication between hams. A ragchew—a contact or an actual conversation over the airwaves—can be between you and someone who lives down the street or someone who lives across the continent and is using advanced technology.

Earlier on, we also mentioned, in passing, nets. A net, short of network, is a predefined meeting of hams on a specific frequency dedicated to specific purposes. For instance, a net is a network that relays subject or topic specific message between operators. As an example, we have emergency nets that meet with the sole purpose of practicing preparedness for actual emergencies. We also have technical service nets that work as forums for ham operation advice and help hams work through the technical details of being ham operators.

About DX-Ing, Contests, and Awards

Meaning distance, DX is a very common term used by many ham radio enthusiasts. The term DX-ing means the practice of attempting contact with the person farthest away from you. For instance, if your antenna has a frequency of 50 miles, DX-ing would be attempting to establish contact with another ham at a 50-mile radius.

When it comes to contest and their awards, the ham world is exciting to the brim with many contests in existence. Most of these contests revolve around specific competition (such as DX-ing) or contacting as many people as possible within a given time. A later section of this guide shall cover contests and their rules.

A precursor to licensing

In this section, we have gotten basic information of ham radios, the culture, and a bit of the history. We shall have a detailed study of the history of ham radio contesting. For now though, before we can move on to the next section (where we'll discuss the cultural history of ham radio contesting), we need to talk a bit about contesting and the license requirements.

To become a ham radio operator, i.e. to be eligible to transmit on an amateur radio frequency, you need a license. This license is easy to enough (well, easy enough).

The test, administered in the U.S. by local volunteers, covers amateur radio rules and regulations as well as electronic theory. Because amateur radios have a vibrant community, you will find tons of study material and guides. The great thing about this license is that it does not have an age restriction (anyone can sit for the test: even children). Other countries and jurisdictions will have their own licensing bodies despite the fact that many countries share the same ham frequencies as those used in the United States.

The license has classes with each license class allowing operations within specific bands. Higher license classes allow the use of more frequency bands. In the U.S., the FCC has recently relaxed the Morse code requirements to make it easier for ham enthusiasts to get an amateur radio operator's license. While we shall have a section dedicated to licensing, the current FCC licensing plan will see you get an amateur radio license if you ace a 35-questions written examination.

Organized in 1914 by H.P. Maxim to help relay radio messages, the American Radio Relay League (ARRL) is without a doubt responsible for helping many enthusiasts get into amateur radio by publishing many study materials and publications. In the U.S., local volunteers who are members of amateur radio clubs are responsible for administering the amateur radio test.

Before we discuss ham radio equipment, let us discuss ham radio contests:

Contesting 101: A Cultural History of Amateur Radio Contesting

As implied earlier, radio sport (also called contesting) is the competitive side of amateur radio. The general aim of the contest is to contact and communicate with as many amateur radio stations as possible within a given time.

That said, contesting, more so the rules and nature of an amateur radio contest, changes from competition to competition with the most defining elements being the amateur bands used, the mode of communication, and the information exchanged during a ragchew. As also indicated earlier, the contacts made contribute to a score that the sponsors publish in a magazine or online publication.

Historically, contesting started developing in the 1920s and 30s from other amateur radio activities. As communication between amateur radio enthusiasts grew and expanded, competitions formed to motivate ham enthusiasts to contact as many amateur radio stations as possible in and outside their countries of residence.

The contest also formed as a way to provide hams an opportunity to practice and highlight their message handling abilities as used for emergency or even routine communications over extended distances. From these contests grew a wide number of contest and today, many amateur radio enthusiasts pursue it as a sporting activity.

Radio enthusiast magazines, radio societies, and radio clubs are the primary benefactors of amateur radio contests. Because sponsors are the organizers of the competitions, they are responsible for publishing the rules of the event. They are also responsible for collecting any operational logs from different stations that are participating in the event, cross checking the correctness of the data in order to generate a score for each of the stations, and publishing the result in a readily accessible platform such as a magazine, website, or society journal.

Most of the competitions—with the exceptions of specific contest for shortwave listeners—are between stations that have amateur radio licenses, and because of this, since most of the radio frequencies used in these competitions prohibit the use of radio frequencies for economic interests, these competitions are informal and there are no professional amateur radio contest or contesters. As such, when you participate in amateur radio competitions, do not expect cash rewards; instead, you will receive trophies, certificates, and perhaps plaques.

The most basic contest is where during a contest, each contest tries to establish two-way communication with other licensed hams on specific frequencies and exchange defined information specific to the nature of the contest.

In most of these contests, the information exchanged varies (its determined by the organizing body). For instance, the ragchew exchange between hams could be signal reports, the location of the station, the R-S-T system, the Maidenhead grid locator, the geographical location of the station, age of the operator, or a serial number.

For each contact made, the amateur radio operator must properly receive the call sign, a unique designation for a transmitter station assigned by the FCC (or any other government agency) or informally adopted by individuals or organizations. He or she must also receive and record in a log specific information defined in the rules of exchange (the information required to be exchanged), and a record of the time of the contact as well as the frequency used to make contact and exchange the information.

The organizing party then takes the log and uses the contacts to tabulate a score for each ham based on the score formula defined by that contest. The most common formula is one that assigns a number of each contact and a multiplier depending on the ragchew or information exchanged.

In contests held in North America on the VHF amateur radio bands, the most commonly used formula is one that assigns a new multiplier for individual (new) Maidenhead grid locator and thus, the competition winner is the one that makes most contacts with other stations in most locations.

Based on the rules of a particular contest, individual multipliers may count only once during the contest or once on each radio irrespective of the radio band that fist earned the multiplier. Because of this, the contest can affix a specific amount of points for each contact or depending on other factors such as the weather, geography and if the communication crossed boundaries such as political or continental.

Other contests award points depending on the points between two stations (the Stew Top Band Distance Challenge is a great example of such a contest). In Europe, most amateur radio contests held on the microwave and VHF award hams 1 point for each kilometer of distance, and therefore, the further the contact station, the higher the points.

After receiving the logs, the contest organizers check them for accuracy. At this point, they are free to deduct points, multiplier, and even credits if they note errors in the contact logs for individual contacts (QSO). As you can guess, the result of the score will largely depend on the scoring formula used. Some scores can be small numbers while others can be large numbers.

Most of ham radio contests in America and Europe have different entry categories. Each category has an individual winner with some contests segmenting winners by geography such as countries, Canadian provinces, continents, and U.S. States. Of these categories, the most common is the single operator category and its many variants.

In the single operator category, one ham operates an amateur radio station throughout the contest. The single operator category further divides into subdivisions depending on the highest power output. In one such division, the QRP category, single operator amateur radio stations must operate the station with no more than 5 watts of output power. The high power category of the single operator category allows hams to run stations that transmit with as much power output as allowed in their licenses.

If you want to get into ham radio as a team, you will have to get into the multi-operator category. This category allows hams to team up to operate a single amateur radio station. Some of the contests in this category ask hams to use one or several radio transmitters used simultaneously.

About Contests

From a contest perspective, different benefactors sponsor a variety of competitions each year with many of them (the sponsors) creating events whose purposes is to promote interest in ham radio as a hobby and requisite skill.

Weekends are the most common contest days. Some contests, however, also use local weeknight evenings. Some competitions are short while others are as long as 48 hours long (for a single session).

As hinted at earlier, each contest will define the stations eligible for participation, the frequencies of operation, the communication modes hams can employ, the frequencies of the amateur radio they should contact, and the time within which to make contact not to mention the information exchange with each contact station.

From this, it's clear to see that the rules of engagement depend on, and change from competition to competition. Some competitions will (and do indeed) restrict the contacts to specific standards such as continent and country (geographical). For instance, the European HF Championship has a sole aim: to foster competition between amateur radio stations in Europe.

Some contests allow amateur radio stations worldwide to participate and contact each other for points. One such contest is the CQ World Wide DX Contest that allows stations to contact other stations located anywhere on the planet. The competition attracts thousands of competition amateur radio stations each year.

Likewise, the amateur radio scene is full of regional contests that allow participation from stations spread across the world with the only restriction being the stations contacted by individual ham radio stations. An example of such a competition is the Japan International DX Competition. In the competition, Japan based amateur radio stations can only contact ham stations outside Japan and vice versa.

Most of the available contests use multiple (or one) amateur radio bands to allow stations in the competition to make two-way contacts. On various occasions, we have mentioned HF-based competitions. These competitions use either (as single or multiple bands) of the following bands; [160 meter](), [80 meter](), [40 meter](), [20 meter](), [15 meter](), and [10 meter](). VHF based contest use bands over 50 MHz.

In some contests, amateur radio stations can transmit on all HF and VHF bands with points awarded for multipliers and contacts on individual bands. Other contests may allow contacts on all bands but restrict the type of contact—perhaps to one contact with individual station irrespective of band, or place a limit on the multiplier to per contest instead of per band.

Amateur radio contests are available for amateur radio enthusiasts from all lifestyles. Some of these contests restrict themselves to CW emissions (meaning they us Morse code as their default mode of communication) while others restrict their communication to spoken communication (meaning they use telephony). Other contests use digital emission modes such as PSK31 or RTTY.

The most common thing in popular contests across the board is that most of them hold the contest on two separate weekends whether they dedicate one weekend to CW and the other to telephony. For instance, the CQ World Wide WPX Contest holds the phone-only contest on a weekend in March and the CW-only contest on a weekend in May.

Contest that restrict contact to single radio frequency bands often let competing amateur radio station use different emissions modes. Most VHF contests normally allow most (or any) emission mode even some digital modes designed to work on specific bands. Because content rules vary, some hams choose to specialize and participate in contests that restrict the use of certain modes.

Because of the nature of contest (the fact that each contest has specific rules) some contests can last up to 48 hours. The most common of such contests are those on the HF bands and contests open to worldwide participation. In most instances, these large (and worldwide) contests start at 0000 UTC on a Saturday morning to 2359 UTC Sunday. As you can imagine, smaller (especially regional ones) contest are shorter in duration with contests that last 4-24 hours being the most common.

Because of the duration of larger contests, some of them employ a concept called "off time" where a station can operate for a portion of the time available. An example of this is the ARRL November Sweepstake. Although the contest is 30 hours long, individual amateur radio station cannot be on air for more than 24 hours. The effect of the "off time" is that it forces competing stations to choose when to be online making contacts and logging them as well as when to be offline (or off air). This element adds strategy to the competition.

In the early 1930s, amateur radio contests could happen on multiple weekends. Although only a small number of contest have held on to this tradition (called cumulative contests), they limit themselves to Microwave frequency bands. To replace these types of contest, nowadays, the most popular types of contests are short "sprint" contest that last a few hours. Some contests such as the North American Sprint Contest increase the excitement and difficulty of the contest by requiring operators to change frequencies after making individual contacts.

As you get started in amateur radio contests, you will be looking for contests that allow newly licensed hams to compete. One such competition is the Maine 2 Meter FM Simplex Challenge. This challenge allows newly licensed hams to participate in contests; it does so by offering entry categories for those with handheld radios (or fully equipped contest stations) and restricting contacts to a singular VHF band.

Because we have a variety of contest, they attract many contestants and amateur radio stations. Because each contest defines its rules, the nature of the contest determines the strategies implemented by competing amateur radio stations to score the most points and multipliers.

Because each contest has specific sponsors who define rules of the contest, there lacks an international authority or governance organization for amateur radio as a sport. As such, there is not uniformity of contest rules. However, most of the contests appeal to all participants to adhere to amateur radio regulations of their individual states or countries.

Brief History of Amateur Radio Contesting

Amateur radio contesting traces back to the 1920 during the Trans-Atlantic Test where radio operators made the first attempted to use short wave radio frequencies to establish long distance communication across the Atlantic Ocean. In 1923, radio operators established the first two-way communication between Europe and North America. At that point, the tests become an annual event where many other stations started establishing two-way contacts over even greater distances.

In 1927, the American Radio relay League, a principle facilitator of the aforementioned tests, put forth a new formula for the annual event and at that point, started encouraging amateur radio stations to make as many two-way contacts with other stations in as many countries as they could.

1929 saw the first amateur radio contest dubbed the International Relay Party, a contest that was an immediate hit (and success). From 1927-1935, the ARRL sponsored the event and in 1936, changed the name of the event to ARRL International DX Contest, a name that

event to date, represents one of the most vibrant amateur radio contests in the world.

Because of the growing interest generated by the popularity DX communication through participation in International Relay Parties, the ARRL proposed and then adopted an event format for non-international contacts.

In 1930, ARRL started the first ARRL All-Sections Sweepstakes Contest, a contest that required a robust exchange of information for all contacts made from two-way contacts, a formula adopted from the National Traffic System (its message header structure).

Because of the robustness of the information exchange, the competition grew in popularity with many operators using it as a way to gauge their operating skills and the robustness of their stations. The event became a primary attraction for those with amateur radio competition in their minds and in 1962, the ARRL sponsored event took on the name ARRL November Sweepstakes.

Field day operating events are some of the most important developments in early amateur radio contesting, with one of the earliest organized field day event held in 1930 in Great Britain, and event later mirrored by many small field day events held across Europe and North America. In July 1933, ARRL organized the first ARRL International Field Day, an event the league popularized through the OST, their growing membership journal.

ARRL promoted field day events as an opportunity for amateur radio enthusiasts to test their emergency and disaster preparedness where they would have to operate from portable locations. Since then, field day events continue to be popular mainly because even though they use the same operating and scoring structures used by other amateur radio contest, they add in an element of disaster or emergency preparedness.

Most modern day amateur radio contests draw heavily on the custom of DX communications, communication readiness, and traffic handling. Since the 1920s, as interest in ham radio activities has grown, so has the amateur radio contest.

In the 1930, radio societies in countries such as Spain, Canada, Australia, and Poland took sponsorship or amateur radio contests. In specific, the 1934, the ARRL sponsored a contest event restricted to the 10-meter amateur radio band. As 1937 dawned, countries such as Zealand, Brazil, Hungary, France, Ireland, Germany, and Great Britain say an influx of sponsored contests.

The first ever VHF contest held was in 1948, a contest dubbed the ARRL VHF sweepstake. On the other hand, the RTTY society of Southern California was the first to organize the first RTTY contest in 1957. The [National Contest Journal](), a publication that started circulating in the United States in 1973, was the first publication specifically dedicated to the sport of amateur radio.

In terms of international contests, the IARU HF World Championship, an event sponsored by the Amateur radio Union and previously called the IARU Radiosport Championship (a name changed in 1986), started in 1977.

In 1986, as amateur radio become popular as a sport, [CQ amateur radio Magazine]() started the Contest Half of Fame. As time went by and the sport's popularity grew, its popularity worldwide grew and tens of thousands of hams started connecting their ham radios, conventions, journals, and websites.

Because the sport does not have a worldwide organizing authority or body, the sport lacks a world ranking system and as such—and because contesters cannot compare themselves to other hams—the challenges faced by competitors depend on locality, proximity to amateur radio communities, and most importantly, the radio propagation rules of the location.

July 1990 was an exciting month because in this month, the world witnessed the first ever "face to face" championship known as World Radiosport Team Championship, an event held in Seattle Washington, an event whose purpose was to overcome some of the challenges presented by previous amateur radio competitions.

The contest saw the participation of top hams from around the globe and required them to participate in the competition from a single location. In the event, 22 teams of 2 persons represented fifteen countries with some from the Soviet Union and the Eastern bloc.

In 1996, San Francisco California played host to another WRTC event; in 2000, Bled, Slovenia played host; in 2002, Helsinki, Finland took the mantle; and in 2006, Florianópolis, Brazil played host to the WRTC. The version of the event that most hams consider the "world championship" took place in 2010 in Moscow Russia.

Learn more about current WRTC events from the resource link below:

http://www.wrtc2018.de/index.php/en/

For a more detailed history of how the technology has developed from the beginning to its present states, read the valuable and insightful resource below:

http://w2pa.net/HRH/

Now that we have a fairer understanding of ham radio contest, the next thing we need to look at is how to get started:

Getting Started: Getting Your Ham Radio License

The last two sections clearly outlined the need for a ham radio license. As the previous sections implied, although the licensing test is simple, at first glance, it may seem confusing and chaotic.

This section of the guide shall walk you through the process of getting your ham radio license because without it, you cannot participate in amateur radio contests. Even when your reasons for involving yourself in ham radio activities may not be contest related, you still need a license if you intend to use your ham radio to send two-way messages.

About FCC Licensing

As implied earlier in this guide, as a US resident, you can get your ham radio license from the Federal Communications Commission (FCC), the federal body governing the ham radio or Amateur Radio Service.

The FCC has three license offerings: (*1*) *the technician license class*, (*2*) *the general license class*, and (*3*) *the extra license class*. The first class, the technician license class, is the introductory license. Before you can earn any of the other licenses, you must earn it.

The main difference among the license classes is the transmit frequency privileges. When the FCC issues a license in any of the classes, it gives the license holder the privilege of transmitting on specific range of frequencies allocated to amateur radio service. The further you climb up the license classes, the broader the frequencies on which you can transmit with the extra license class allowing a license holder full use of all frequencies dedicated to amateur radio services.

Licensing: About Frequency Privileges

As you can probably note from the above discussions, the different license classes offer different transmission frequencies. Each of the licenses and frequencies dictates the type of on air-operations you can conduct. It is, therefore, very important to understand the various license classes because that understanding will help get you off to a great start as a ham radio enthusiast.

Before we can delve into the various licenses classes and what each allows you to transmit, we need to backtrack a bit and cover basic radio concepts that will help you understand the differences in license classes and their in-built on air operation.

About Frequency Bands

A frequency band is a contiguous range of radio frequencies. Simply put, radio signals, which are what your ham radio sends out and receives, utilize magnetic and electric fields to hop from antenna to antenna. The frequency is the rate at which a signal travels back and forth.

Licensees of the technician class license use radio frequencies expressed in millions of back and forth cycles per second (or in some cases, hundreds of millions of cycles per second). The unit of measuring the cycle per second is hertz and that of millions of cycles/second is megahertz (MHz), something we have mentioned in passing many times in this guide.

To simplify this concept even further, consider a signal that wiggles from 144 MHz to 148 MHz on a frequency band. From a nontechnical perspective, your FM radio has a commercial frequency band of 88 MHz to 108MHz; you can use your tuner to access any of the radio stations transmitting within these frequencies.

Just as the FCC has allocated specific frequency bands to commercial FM radios, it has allocated such frequency bands to Amateur Radio Services; these frequencies are what we call ham bands; each frequency band within the ham band will have a different propagation or traveling characteristic.

About Operating Mode

The AM and FM bands on your car or home radio are two different operating modes: the amplitude modulation (AM) and the frequency modulation (FM). These two modes are different in that they use different encoding mechanisms to transmit information into radio signals.

Learn more about AM and FM operating mode from the following resource:

https://www.diffen.com/difference/AM_vs_FM

Ham radios use AM and FM modes as well as other modes such as Morse code and computer-generated digital packets among many other digital modes. In essence, a mode is the system used to transmit information to and from a radio wave.

Now that you understand this, let us look at license classes and the various things each license class allows you to do:

License Classes and their Various Transmission Privileges

Earlier, we introduced three license classes:

(1) The general license class

(2) The technician license class

(3) The extra license class

As we go through the license classes, use the [US Amateur Radio Bands Chart](#) as a reference:

Technician license

As implied earlier, this is the entry-level license. Once you get this license, you immediately gain access to amateur radio frequency bands of up to 30 million cycles/second and higher, radio frequencies normally regarded as VHF (Very High Frequency), UHF (Ultra High Frequencies) and higher microwave frequencies. With access to these bands, you can use them for voice modes such as FM and AM and even for digital modes such as Morse code.

Normally, VHF frequencies (and other high frequencies) travel further than line-of-sight and are thus what we call locally propagating signals. On these frequencies, if you have a powerful antenna placed on a high mast, you can transmit and receive signals over many miles with the distance depending on the terrain of the area. As we also implied earlier, many areas have repeaters that allow for the extension of the range of communication.

With a technician license class, you can participate in local nets or on air meetings that involve several hams and in cases where the hams you connect with are using repeaters, you can communicate with hams in different states (of course depending on the terrain and your equipment range). To increase the distance of communication, some hams use their VHF and UHF privileges and satellite repeaters orbiting earth.

A technician class license limits your privileges on high frequency bands (bands in HF, 3 to 30 MHz). As you know, there is a difference in propagation of the different frequencies. For instance, HF frequencies will propagate differently from VHF or UHF.

On their part, high frequencies can reflect on the ionosphere, a layer of earth's atmosphere, and thus travel further, even around the globe. The technician license allows you to transmit voice on a HF band of 10-meter band and Morse code on 4HF bands. Essentially, this means that with this license, you can contact other hams in different countries across the world.

General License

As stated earlier, to earn this license, you need to earn the technician one; this license simply expands the privileges of the technician license by allowing you access to additional operating modes and higher frequency bands.

When you earn this license, you can operate on all amateur bands allocated to amateur radio services by the FCC. You can use voice modes on all allowed HF bands (7 HF bands), which allows you to make voice contacts over greater distances—again, using the ionospheric skip propagation—and can also use Morse code and other digital modes in 8 HF bands that propagate around the world.

Extra License

This license expands the privileges of the general class. It allows the use of extra class exclusive band segments and essentially, gives you access higher frequencies whose propagation exceeds the local area. Because this class offers more communication possibilities, it requires more skills and technical knowhow.

For the remainder of this section, we shall restrict our discussion to the technician class license because although it's an entry level license, it offers a ton of frequencies and communication capabilities.

About The License Examination

To earn any of the three licenses, you have to pass an examination. Each of the classes has an examination that tests your grasp of amateur radio rules, operating procedures, regulations, and other technical topics.

The questions for each license come from a pool of exam questions with each question revised every 4 years. The questions for the latest technician class license come from a questions pool created in July 1, 2014 and that will remain valid until June 30, 2018. The questions are often simple and the public domain for the development of study guides and material.

The current entry-level class license has a pool of 426 multiple questions. However, the exam itself has 35 questions selected from the pool and where after reading the questions, you choose option A, B, C, or D as the correct answer to the questions. To pass the technician exam, you must score a pass rate of 74% (or get right 26 out of the 35 questions).

In modern day, these exams do not require a demonstrated ability to communicate in Morse code and thus, to get a license—whichever license class—all you have to do is pass the multiple-choice question exam.

The National Conference of Volunteer Examiner Coordinators (NCVEC) is the body responsible for formulating the questions pool and their volunteer examiners (VEs) are the ones responsible for administering the exam across the United States. These volunteer VEs are ham enthusiasts certified and sanctioned by the FCC.

Where To Find An Examination Session

If you live in a populated locality that has a vibrant ham radio community, you will find several exam sessions held each month at ham radio events of held by ham radio clubs. If you live in a densely populated area that lacks a vibrant ham radio community, you may have to travel a short distance to attend an examination.

Use the following resources to search for scheduled VE exams:

http://www.arrl.org/find-an-amateur-radio-license-exam-session

http://www.w5yi.org/exam_locations_ama.php

Most VE administered exams allow walk ins; however, some will require an advance registration.

What To Bring To The Exam Room

When going to sit for the exam, you will need to carry with you, a driver's license (or any other state issued photo ID such as your social security card—if you have one—student ID, birth certificate, utility bill, and the likes). You will also need to carry money ($15) for the fee payment, a calculator, and pencils.

How to Prepare for the License Exam

How you prepare for the exam will determine how great you manage. Here, you have various options: you can memorize enough questions to pass the exam, which is not a great strategy because even though you may ace the exam, you will not have a clear or firm understanding of the concepts and radio terminologies.

Learning and understanding the ham radio and the concepts behind it is the best study strategy because it allows you to understand what the exam questions are asking of you, which will allow you to figure out the right answers.

To study for the exam (and learn more about ham radios), you can use study guides or attend a class offered by ham radio clubs. You can find study guides on the resources below:

http://www.kb6nu.com/study-guides/

https://hamstudy.org/

http://www.aclog.com/aprs/study.php

Amateur Radio Competition Equipment

Now that you have your technician class license, to get started, the other thing you need to do (if you have not) is to set up your equipment and station. At the most basic level, you will need a ham radio that can transmit and receive signals using the various modes.

In this section, we shall discuss the various equipment you need to participate in ham radio contests. We shall start with the basics:

Transceivers

To get started immediately, this guide recommends that you get an FM transceiver. Because the technician class offers you access to VHF and UHF frequencies and repeaters on which is the most popular voice mode, FM, you can use an FM transceiver to send noise-free voice messages in a trouble free manner.

Most available handheld transceivers are FM only, so are most of the transceivers in cars—mobile operations—or the ones used to create a base station. Most of these transceivers restrict themselves to 1-3 VHF-UHF amateur bands.

Another option here is a multimode transceiver. These types of transceivers are a bit pricey but offer complex operations and capabilities. To transmit voice over HF bands, you will need a transceiver that has a single sideband mode. This mode is very special in that it optimizes use of power and is the most popular for long distance communications on HF bands as well as over-the-horizon voice transmission.

A multimode transceiver is ideal for when you upgrade your license to the general class license because the license expands the phone mode privileges on the HF bands. Further, the transceiver (a multimode transceiver) allows you to communicate in Morse code as well as attach a computer for transmissions on digital modes.

Now that you have your technician class license, get a radio. The most commonly used transceivers are the handheld HTs.

Commonly called HTs or handy-talkies by most hams, you can start out by buying a FM HT that offers 2-meter band on VHF and 70-centimeter band on UHF. These are the most commonly used handheld transceivers since most available repeaters use these frequency ranges. With that said, some parts of the United States use the 1.25-meter band for repeater operations.

While you can start with the most basic HT (which transceiver you start out with shall depend on your level of knowledge and expertise), you have many radio choices with many radios offering various inclusions and features. For starters—especially if you are relatively new to ham radios—go for a HT radio that does not appear too complicated (they are easier to manipulate and are not as expensive). If you are unsure of which HT to buy to get started, join a club or seek the mentorship of an Elmer (an experienced ham radio operator).

Learn more about buying transceivers from the following resources:

http://rsgb

http://rsgb

In closing this transceiver section though, we have to mention that purchasing radios is perhaps the most costly aspect of setting up an amateur radio station primarily because the radio is very central to ham radio activities. This therefore means that choosing the right transceiver/radio is very important (which is why this guide suggests that if you are unsure of which radio to buy, to seek assistance from an Elmer).

Now that you have chosen a radio, you also need to pay some attention to filters. A filter allows you full use of a desired radio signal while reducing the signal strength of other signals in your locality.

Now that you're setting up your amateur radio station, you will need a good filter because an effective one makes ham radio operations and communications easier. Whether you need to buy a filter will depend on the type of transceiver you choose to use since radios offer cascading filters that follow one another while others provide extra filters.

In addition to a transceiver, you may need one or any of the following (depending on your chosen transceiver).

Power Supply

If you opt to start out on a handheld transceiver, one you can use on UHF (FM or VHF, you will need rechargeable batteries (read the device manual to know the type of power supply you need for your handheld transceiver).

If you decide that you'll set up the transceiver at a specific area of your home, you shall need to invest in a DC power supply whose voltage shall depend on the voltage needs of your transceiver (most handheld transceivers have a voltage requirement of 13.8 volts and you may need to buy an adapter.

If you have a base station—a permanent or mobile one—you will need a power supply of no less than 13.8 volts (check the device manual for voltage specifications). Some base stations will come equipped with a 230 volts power supply inbuilt. Most transceivers will have a manual telling you the maximum voltage for individual devices so that the power supply you get is adequate.

The types of power supplies in the market are switched-mode and linear. The linear type has a bulky design that uses two transformers to change the input voltage of 230 volts into 13.8 DC power for use by the base station. Linear power supplies are heavy and large.

The switched-mode ones are vastly different in that they directly convert the AC voltage into DC and then filter it. The High voltage DC is subsequently fed into a power oscillator, which usually switches it on or off at a frequency of 20 to 50 kHz with the result being a pulsating DC converted to 13.8 volts by a transformer. Switched mode power supplies are smaller, lighter, and cheaper than linear power supplies. However, even though these power supplies are smaller and cheaper, take note that in some instances, they may interfere with signals transmitted over your radio. To ensure zero signal interference, choose a switched-mode power supply that has a low radio frequency interference (RFI) feature or one that has a knob you can use to adjust interference when you note it.

Coax

A coax shall help you connect your antenna and transceiver. Since most radios have a 50 Ohm unbalanced output, you will need a 50-Ohm coax.

Before you buy a coax, you should give proper thought to the quality of the coax you intend to buy and its characteristic losses. In this regard, note that the higher the frequency, the greater the coax loss. For operating on lower HF bands, a 5mm RG58 coax will suffice. When you move to the higher high frequency bands of 24-28 MHz or 144-430 MHz, you will need a better coax.

The most commonly used coax is the RG213 because other than offering lower losses, it's also less flexible and thicker.

Antennas and Antenna Analyser/SWR meter

The antenna you choose to use for your base station can be directional or omnidirectional. Directional antennas beam concentrated signals in one direction while omnidirectional ones radiate the signal out equally. If you have a mobile transceiver, you'll note that it uses whip antennas you can interchange for different bands.

Most modern transceivers have inbuilt SWR meters that help with the setting up of an antenna. Additionally, many external antennas units have inbuilt SWR meters that make finding matches easier.

If you are assembling a base station in a car, you will need an SWR meter so you can correctly set up the antenna. The same case applies to a home base station; when choosing which one to buy, consider the bands you will be transmitting on.

Further to this, if you will be experimenting with different types of antennas, you will need an analyser, which although not cheap, will be very essential and offer more information—such as faults in cables, measure of the cable electrical length, and a graphical display of the SWR curve—than an SWR ever could.

The first antenna you set up should be a half-wave dipole antenna.

To learn more about setting up your first antenna, read the content on the invaluable resource below:

http://rsgb.org/main/get-started-in-amateur-radio/antennas/your-first-antenna-the-half-wave-dipole/

Log Book

Since you will be participating in amateur radio contest, and while it is not a necessity, you will need a log book where you can log your contacts. The most inexpensive way to keep a log is to have a paper log. However, for contest purposes (which is the purpose of this guide), it's important to have a computer log. Fortunately, there're plenty of software (for both Mac and PCs) with some being paid and some free.

While a paper log is easy to keep, an electronic one is ideal for competitions because it automatically logs all you contacts and the ragchew exchange between you and the other station. To add to this, most of the available logging software also print labels for your QSL cards, track your progress towards any goals (if you're in a contest), and also upload the log or information to the ARRL's logbook.

To learn more about logging software, see the content on the resources below:

https://en.wikipedia.org/wiki/List_of_amateur_radio_software

http://www.w1wc.com/software/

https://www.dxzone.com/5-free-ham-radio-logbook-programs/

To learn more about the equipment you need for ham radio competition, as well as the various logging software available, read page 5-15 of this free contesting manual:

http://k4ro.net/w4phs/W4PHS_Guide_to_Ham_Radio_Contests.pdf

Making Your First QSO: A Simple, Ham Radio Contest Guide to Making Contacts

Now that you've set up your base station and you're ready to start contacting other station, you're ready to make your first contact (your first QSO). Before making your first on-air radio contact, this book recommends that you listen in to a few conversations. If you're in the vicinity of a repeater, monitor it and then tune it to the ragchew exchange between the other hams to note the type of conversation going on, the language used, and such. Most ragchew exchanges will use common language with a few ham radio specific terms thrown in there.

The easiest way to learn the lingo is by looking over the resource below that has a list of common ham radio terms:

http://www.hamradioschool.com/wp-content/uploads/2014/04/common_ham_radio_terms.pdf

After listening in for a while and when you feel ready to make your first contact, use the knowledge you used from your exam study guide to make the first on-air contact by pushing-to-talk and using your new call sign—

assigned to you after passing your technician license test.

While your first QSO can be daunting—not to mention the anxiety that comes with contacting experienced hams when your just a newt—you can arrange to test the waters by arranging your first QSO with a friend, perhaps one from the club, or an Elmer if you have one. This will make the prospect less stressful and instead fill it with fun.

In addition, this guide recommends that you get started by making your first QSO on a 2m FM instead of a HF SSB because the latter can be noisier and prone to interference.

Before you get started—this applies whether you're contacting other hams for fun or are in a contest—you need to make sure your radio is on the right mode, you have the right amount of power, and that the microphone gain is correct. Make sure, also, that you're using the correct antenna that matches your needs and radio. Here, you can use an ATU or feed the line directly into your radio. For the latter, ensure that the antenna has a low SWR.

If you intend to make a phone or voice QSO, you will have to (1) call CQ or (2) answer someone who is actually calling CQ. What does

CQ mean? Well, a CQ is a general call targeted at no one in particular and is the traditional ham radio way of reaching out to new ham radio contacts. Obviously, before you make or receive a CQ call, you need to find a frequency not in use by another station, which is usually not easy especially if you live in a ham radio vibrant locality where there's crowding on the HF bands.

About Finding An Open Frequency

Finding an open frequency is a process. The first thing you need to do is tune your antenna tuner to as close as 1:1 as you can and then check to see if someone is using the frequency. Not hearing something is not an indicator of a frequency not in use. Someone could be using a frequency, say nothing, but still hear you.

To check if a frequency is in use, send "QRL" (a term used to ask if the frequency is in use or if the ham on the other side is busy) and if you're on voice, simply state, *"This is (insert your call sign here). Is this frequency busy?"* Depending on the reply you get, either move on or use the frequency if you don't hear a response.

The key idea here is to listen before talking or transmitting.

To call CQ, use the 3 X 3 method. In the 3 X 3 method, you call CQ three times in the format below:

"CQ CQ CQ this is Dave Six Alpha Bravo Charlie, Dave Six Alpha Bravo Charlie, Dave Six Alpha Bravo Charlie standing by."

When you use this formula, a station may come back to you and say:

"N1ABC (or whatever) N1ABC, this is Dave Six Alpha Bravo Charlie. Good evening/day, your report is 59 (or whatever it is), my name is Mike – Delta Alpha Victor Echo – and my QTH is London – Lima Oscar November Delta Oscar November."

NOTE: It's important to point out here that the informational exchange will vary from contest to contest. Some contacts will come back to you with your report, name, and QTH while others will not. For competition purposes, its important to note the kind of information needed in the exchange so that if a station does not offer that information, you can ask questions to get it.

Other than taking note of the above as it relates to competition, what you talk about with your new contact will vary. For instance, you can exchange information about your stations, on whatever else you want.

The point of note here is that you should conduct yourself with decorum, as if anyone in the world is tuning into your conversation, which is not a too farfetched notion because in the real sense, anyone could be listening in. As such, avoid conversations around hot topic that draw out the worst in people—politics and religion are prime examples of topics to avoid or terminate when they start turning into arguments.

In some instances, especially if you've established a QSO contact with a station operated by a non-English speaking person, the other ham will wish you a "73" and move on. 73 is a Q code for best wishes and in this case, it simply means the person operating the station is non-English speaking and is wishing you well because he or she does not want to get tongue tied. 73 is also an ideal way to end a conversation.

Check the resource below for a full list of QR codes:

http://www.amateur-radio-wiki.net/index.php?title=Codes_and_Alphabets

Please note that for contest, you need to understand the contest rules so you can know which information to exchange when you make your QSO.

After making your first QSO and accustoming to operating your ham radio, using repeaters and radio bands, and contacting other hams directly without using repeaters (called talking to other ham simplex), you can then consider upgrading to something that has a higher power output than the handheld. The more you immerse yourself in the ham universe, the faster and better your skills shall grow, and the more effective you shall be when competing. You will also have a ton of fun.

Now that you know as much as the next ham, the other thing we shall talk about are the various types of contests, their rules, and past winners.

The Main Amateur Radio Contests and Their Rules

Now that you're immersing yourself in the ham universe—not to mention getting better at making contacts and operating your station (practice makes perfect)—you will be looking to get into the exciting world of amateur radio contesting.

The great thing about amateur radio is that it has a vibrant community and because of this, you will never lack a contest to participate in; in fact, at any given moment, you may find a contest underway in almost any part of the world.

In this section of the guide, we shall discuss the main amateur radio contests held during the year, when they happen, their rules, and past winners.

ARRL DX Contest

The purpose/mission of this contest it to encourage stations to enhance their knowledge of DX propagation on HF and MF bands and improve their operating skills. The contest also seeks to improve hams and station capabilities because the completion seeks to have DX stations only contact W/VE stations. W/VE stations are stations operating in the United States and the District of Columbia (minus Hawaii and Alaska), and most Canadian territories and provinces except those on Sable and St. Paul Island.

The contest defines DX stations as any non-W/VE station including those in the US and territories such as the Pacific and Caribbean. It defines DXCC entities as those defined by the [ARRL DXCC list](). The contest demands an electronic log submitted by email or a memory device. It defines an automated multi-channel decoder—such as CW skimmer—as a device that reveals information about the frequency and identity of the contest station without the direct control and participation of operating ham.

The contest happens on:

CW: the 3rd full weekend in February (for instance, February 17-18, 2018)

Phone: The 1st full weekend in March (for example, March 3-4, 2018)

The entry categories for the contest are single operator, single operator unlimited, single operator single band, and multioperator.

The required informational exchange between contacts and stations has to be:

W/VE stations send signal reports and province or state while DX stations send signal reports as well as a number showing transmitter output power.

When it comes to scoring, points for contacts (QSO) are three for W/VE stations per DX QSO and three for W/VE QSO. For the multipliers, for the W/VE stations, the points depend on the sum of DXCC entities per band (except for those located in U.S. and Canada) while for DX stations, the contest scoring rules states this:

"Sum of US states (except KH6/KL7), District of Columbia (DC), and Canadian provinces/territories: NB (VE1, 9), NS (VE1), QC (VE2), ON (VE3), MB (VE4), SK (VE5), AB (VE6), BC (VE7), NT (VE8), NF (VO1 – see note, LB (VO2 – see note), NU (VYØ), YT (VY1), PE (VY2) worked per band (maximum of 63 per band). Note – although VO1 and VO2 have been officially merged into a single province, they are counted separately in this contest for reasons of tradition."

The final score is a sum total of the QSO points multiplied by a total of the multipliers.

To learn more about this contest, the general rules of engagement, as well as previous and upcoming events, check out the resources below:

Arrl

Rules for arrl contests

Rules for arrl contests below 30-MHz

Rules for arrl contests above 50-mhz

You can see records of past winners from the resource below:

http://www.arrl.org/contest-records

CQ World Wide DX Contest

Held on the last full weekend of October for the SSB and the last full weekend of November for the CW, the aim of this contest is for amateur ham enthusiasts to contact as many stations as they can in all CQ zones and countries (or as many as they can).

The contest allows the use of 6 bands: 1.8, 3.5, 7, 14, 21 and 28 MHz and encourages contestants to observe the established bands. For the exchange, the SSB contest asks contestants to exchange RS report and the CQ zone number of the station contacted; for the CW contest, contestants must exchange the RST report and the CQ zone.

When it comes to scoring, the final score is the total contact (QSO) points multiplied by sum of zone and location/country multipliers. To earn QSO points, contesters can only contact a station once on each band and normally depend on the location of the station contacted.

For contacting a station in a different continent, hams earn 3 points. For contacting a station within the same continental bounds, but in a different country, a station earns 1 point with the exception being the award of 2 points for contacting stations in different countries within the Northern American boundaries. Contacting stations within the same country does not earn you QSO points but count as zone and country multiplier credit.

The contest has two types of multipliers. The first one is an application of a multiplier of 1 for each contact made on each band for different CQ zones. The second is a country multiplier of 1 for different stations contacted from different countries on each band.

The contest allows the following entry categories; single operators, single operator assisted, QRP assisted, classic operator, rookie, and multi-operators.

You can learn more about this contest, rules and regulations, future contest dates, as well as past winners from the resources below:

https://www.cqww.com/rules.htm

https://www.cqwpx.com/

https://www.cqwpx.com/score_db.htm

ARRL "Field Day"

Held on the last weekend in June, the objective of this contest is for amateur radio stations to contact as many amateur radio stations on all amateur radio bands excluding those on the 12, 17, 30 and z60-meter bands. The secondary aim is for hams to learn how to operate their stations in non-optimal conditions.

The competition is open to all participants in countries within in the IARU region 2 and all areas covered by the ARRL/RAC field organization. You can contact other stations outside these areas for credit.

You can learn more about this field day event from the following resource:

http://www.arrl.org/field-day

To know which contest is happening when—not to mention the rules for each contest—use the resource below:

http://www.contestcalendar.com//contestcal.html

The following resource also has a free guide that outlines various aspects of contesting and rules:

http://k4ro.net/w4phs/W4PHS_Guide_to_Ham_Radio_Contests.pdf

In closing this section, it's important to point out that competitions you can engage in are many; each will have a different set of rules. Read a summary of the rules for the content you intend to enter and then download a complete version of the rules so you can assess them.

In addition to following contest specific rules, take note and adhere to FCC rules as well as good operating practice and ham radio etiquette.

Contesting Tips and Techniques for Better Amateur Radio Contesting

Tips to help you improve your ham radio expertise are many. In this section, we shall discuss specific tips that applied, shall help you win more contests (and enjoy them more).

Look Over The Contest Details Beforehand

This is perhaps the most important tip in this book. You don't want to go into a contest blind, which is why you should look over the contest details, rules, times, and other such things before you go about entering into a contest. The most important thing to check is the time of the contest—given that the have different time zones, synchronize the time of the contest with your time zone—as well as the rules of the contest. You also need to understand the information you need to submit on the log because they change from contest to contest.

It also a good idea to look over past contest winners to understand the top scores for the previous year. This will give you an idea of the scores to aim for as you participate in the competition as well as the scores that count as credible.

Test Your Equipment

This is just as important as the tip above. Because competitions are fierce, you should test your equipment before entering into a competition to make sure everything is in proper working order. Particularly, check out the antenna system (and if need be, install a new one or improve upon your current one).

Check the antenna for corrosion—antennas left outside have a tendency to wear and because of it, lose their effectiveness—and especially check the joints and connections.

To earn more points in contests:

Always Go For Multipliers First

While QSO points are important, those that add to the multiplier count have more weight. Because of this, as you compete, specifically seek out, and contact multiplier stations (of course depending on the contest rules) before contacting other types of stations. Some loggers such as the N1MM will highlight multipliers in red from the spotting list and have a separate screen that shows multipliers available on a band.

Be Mindful Of Where You Tune

When checking through your spotting list of stations, take care not to go outside the frequencies defined by the contest rules and your licensed frequency range. For instance, if you're competing in the general class ticket, take care not to contact stations in the extra class region and if you have an extra class ticket, take care not to go outside the U.S. frequency bands.

As a general class operator, to avoid extra and advanced regions that are on the lower end of the bands, start at the lowest, legally acceptable frequency in this class band and then work your way up towards the higher frequencies.

Work As Many Stations In Similar Directions

If you're using a directional antenna, an effective strategy is to point the antenna in one direction that shows multiple stations, work down the spotting list (of course the stations you contact shall depend on the contest rules) and contact as many stations as possible.

Once done with that block of stations, swing the antenna in another direction and repeat the process. For instance, if you point your antenna towards China, work all the stations in that region, then swing it towards Europe, and repeat the process.

Work Through Searching And Pouncing

If you implement the tip above, you should have a healthy list of contacts to work through. When your list dries up, don't spend most of your time calling CQ. Instead, search, pounce, and then when you have a healthy list, run again.

Breaks Are Important

As we indicated earlier, some contests can last as long as 48 hours. While participating in such contests, it's important to take occasional refreshing breaks. Being too tired will not help you win a contest; instead, it will take the fun out of contesting.

Conclusion

We have come to the end of the book. Thank you for reading and congratulations for reading until the end.

From everything we have discussed in this guide, you can see that ham radios are fun (and a very useful skill to have) and engaging in contest is but a fun way to enhance your amateur radio knowledge and expertise.

Once you work through the formative parts of your entry into the wonderful world of amateur radio, work your way into contesting as a way to challenge yourself and enhance your operational and technical knowledge as well as your ability to make contacts in different situations and scenarios.

If you found the book valuable, can you recommend it to others? One way to do that is to post a review on Amazon.

Click here to leave a review for this book on Amazon!

Thank you and good luck!